A Brief Guide to CCC

China Compulsory Certification

Julian Busch

Copyright © 2013 Julian Busch

All rights reserved.

ISBN: 1484115538
ISBN-13: 978-1484115534

CONTENTS

Chapter One

Basic Knowledge about CCC

1. The Origins of CCC

The China Compulsory Certification, also known as CCC or “3C”, is the mandatory qualification system used by the People’s Republic of China to evaluate whether national standards and technical regulations are met in order to protect the health and safety of its citizens, the environment and national security. Certification is compulsory for all products imported into China, sold in China or used in China.

The CCC certification system came into effect in 2002 according to the legal requirements and guidelines of the People’s Republic of China.
Once a product is subject to CCC, the certification is compulsory, regardless whether it is imported to China or produced in China. Products which need CCC are only allowed to be imported to China, sold in China and to be used in China if the products have obtained a CCC certificate.

Timeline of the development of the CCC certification system

- 1978: China becomes a member of the ISO
- 1984: Establishment of CCEE
- 1989: The first catalogue for CCIB certification is released
- April 2001: AQSIQ and CNCA are both founded
- December 2001: China becomes a member of the WTO
- December 2001: The first administrative documents regulating CCC are released
- May 2002: The CCC system is brought into effect

The establishment of the China Compulsory Certification system was undertaken by the Chinese government in order to fulfill its commitments to the World Trade Organization (WTO, http://www.wto.org)[1]. In accordance with the WTO's principle of "national treatment", the CCC system was put in place in order to remedy the inconsistencies between certification of domestic and imported products thus implementing uniform catalogues, standards, technical specifications and qualification evaluating processes as well as uniform marking and pricing standards.

Since China's membership into the International Organization of Standardization (ISO, http://www.iso.org), it has begun to establish and develop product certification systems according to international standards. However, at the time of the country's induction into

[1] http://www.people.com.cn/GB/jinji/31/179/20011207/621727.html

membership, there was no single administrative body to oversee certification, thus multiple standards were applied to imported and domestic products.

For example, the former Administration of Quality and Technology Supervision of the People's Republic of China oversaw the implementation of CCEE (the China Commission for Conformity Certification of Electrical Equipment) certification for domestic products. Whereas, the former Entry-Exit Inspection and Quarantine Bureau of the People's Republic of China was in charge of the implementation of CCIB (the China Commodity Inspection Bureau) certification for imported products.

In April 2001, to fulfill its commitment to the uniform treatment of product certification, the State Council made a decision to merge the Entry-Exit Inspection and Quarantine Bureau of the People's Republic of China and the Administration of Quality and Technology Supervision of the People's Republic of China, into the General Administration of Quality Supervision, Inspection and Quarantine of the People's Republic of China (AQSIQ). In the meantime, the State Council decided to establish the Certification and Accreditation Administration of the People's Republic of China (CNCA) to oversee the nation's product certification processes.

In December 2001, AQSIQ, together with CNCA, issued the Regulations for Compulsory Products Certification and a series of documents including the first catalogue of Compulsory Products Certification, which were the first administrative and regulative documents of CCC. These publications stated that, starting 1 May 2002, CCC certification would be implemented for 132 product classes under the organization and administration of CNCA. Additionally, CCC certification was then officially introduced to replace CCEE and CCIB certifications.

Legal Framework of the CCC Certification System

The legal framework of the CCC certification system includes[2]:

- The Standardization Law of the People's Republic of China,
- The Products Quality Law of the People's Republic of China,
- The Inspection Law for Import & Export Commodities of the People's Republic of China
- The Management Rule of Quality Certification for Products of the People's Republic of China,
- The Implementation Rule of the Inspection Law for Import & Export Commodities of the People's Republic of China,
- As well as other relevant regulations.

2. The Certification Process

For any product subject to CCC certification, as specified in the Catalogue[3], the applicant—manufacturer, seller or importer of the

[2] http://www.cccap.org.cn/eccap/ApplicationGuide/ApplicationGuide.jsp

product—will need to submit an application to the designated certification body approved by CNCA.

The certification process generally involves six steps. However, each authorized certification body (e.g. CQC, CCAP) may have a different application procedure.

Basic certification overview

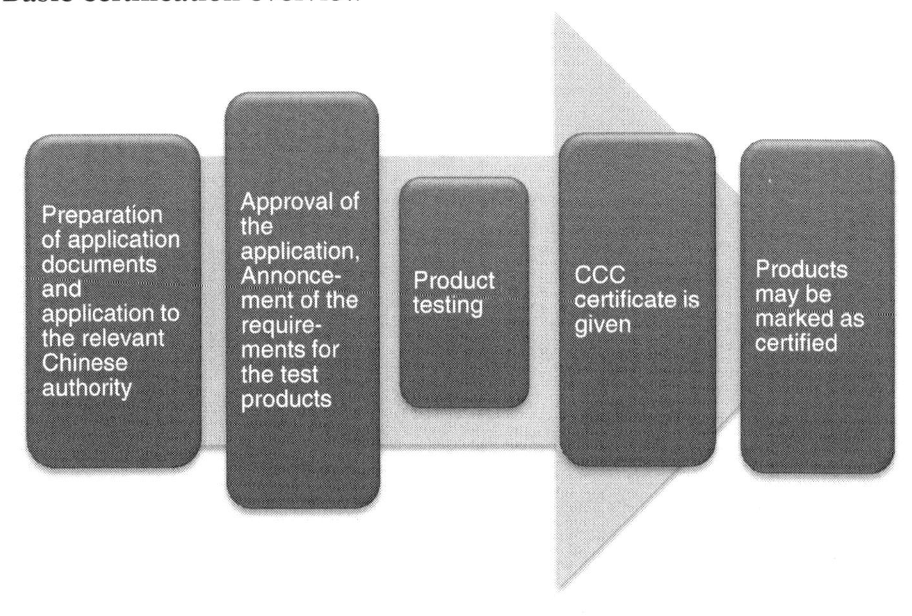

[3] Article 13, Regulations for Compulsory Products Certification, http://www.cnca.gov.cn/cnca/rdht/qzxcprz/flfg/72303.shtml

Process Overview

(1) Application and acceptance

An application and supporting materials are submitted.

(2) Product tests

A CNCA-designated test laboratory in China tests product samples.

(3) Factory Inspection

The certification authority sends auditors to inspect the manufacturing plant. These audits usually take two days.

(4) Evaluation of the Results

The results of the product tests and factory inspection will be reviewed.

(5) Approval of the CCC Certificate

If the application is successful, a Certificate will be assigned. If the application is not successful, the CCC review will contain a detailed explanation. Retesting and further inspections are possible.

(6) Product Marking

After approval is given by CNCA, (Permission of Printing), products can be marked with the CCC-mark.

(7) Follow-up Certification and Audit

Follow-up certifications will be conducted every 12 months after the approval of the first certificate. The procedures are very similar to the first certification but the mandatory follow-up audit usually only takes one day. The results of the follow-up-inspection will be checked and the CCC Certificate will be confirmed if all requirements are fulfilled.

Products must comply with the Chinese GB-Standards at all times. Chinese regulations require that at least one test is done by the factory per year in accordance with these standards. Tests can be conducted in the factory itself or a local test lab. Every five years, a factory will usually receive an updated version of the CCC certificate.

Timing

The time needed for each step of the process can vary greatly based on the plant's familiarity with CCC procedures, visa procedures for the Chinese auditors as well as the planning of the certification authority. As a general guide, the following timeline gives a basic overview of the time needed for each step of the process.

Timeline	Step
Week 1-2	Preparation of the official Chinese application documents and supporting materials (e.g. business license, ISO certificate etc.)
Week 3	Submission of application
Week 5	Receipt of the official test sample request from the certification authority
Week 7	Shipment of test samples to China
Week 15	Factory inspection
Week 19	Receipt of CCC certificate and test reports
Week 20	Submission of application for permission to print CCC mark on the certified products
Week 23	Receipt of the Permission of Printing

3. Products Requiring Certification

All products listed in the Catalogue published by AQSIQ and CNCA require a certification. In addition to the Catalogue, CNCA and AQSIQ regularly announce new products that are subject to certification as well as products that are no longer subject to CCC certification.

CNCA Announcements specifying the range of products subject to CCC certification[4]:

- AQSIQ, CNCA Announcement [2001] No. 33, The First Product Catalogue Subject to Compulsory Product Certification
- AQSIQ, CNCA Announcement [2004] No. 6, The Catalogue of Decorating and Refurnishing Products Subject to Compulsory Product Certification
- AQSIQ, CNCA Announcement [2004] No. 62, The Catalogue of Safety Protection Products Subject to Compulsory Product Certification
- AQSIQ, CNCA Announcement [2005] No. 137, The Catalogue of Motor Vehicle Parts Subject to Compulsory Product Certification
- AQSIQ, CNCA Announcement [2005] No. 198, The Catalogue of Toy Products Subject to Compulsory Product Certification
- AQSIQ, CNCA Announcement [2006] No. 103
- CNCA Announcement [2007] No. 9, The Table of Description and Specification of the Catalogue of Compulsory Product Certification

[4] Catalogue Description and Specification, http://www.cnca.gov.cn/cnca/rdht/qzxcprz/rzml/default.shtml

- AQSIQ, CNCA Announcement [2008] No. 7, Announcement on the Implementation of Compulsory Product Certification on Several Information Security Products
- CNCA Announcement [2008] No. 25, Announcement on the Cease of Implementation of Compulsory Product Certification on Fiscal Cash Registers
- AQSIQ, MOF, CNCA Announcement [2009] No. 33, Announcement on the Adjustment of Implementation Requirements of Compulsory Product Certification for Information Security Products
- CNCA Announcement [2010] No. 26, Announcement on the Implementation Requirements of Compulsory Product Certification for Information Security Products
- AQSIQ, MPS, CNCA Announcement [2011] No. 55, Announcement on the Implementation of CCC Certification on Several Fire Fighting Products
- AQSIQ, CNCA Announcement [2012] No.117, Announcement on the Cease of Implementation of Compulsory Product Certification Administration for Several Products

Overview of Categories

The list below offers an overview of the top categories found in the Catalogue, which are generally subject to CCC certification[5]. Nevertheless, to ensure the necessity of CCC certification, please refer to the original list and announcements mentioned above.

1. Electrical Wires and Cables
2. Circuit Switches, Electric Devices for Protection or Connection
3. Low-Voltage Electrical Apparatus
4. Low Power Motors
5. Electric Tools
6. Welding Machines
7. Household and Similar Electrical Appliances
8. Audio and Video Apparatus (not including the audio apparatus for broadcasting service and automobiles)
9. Information Technology Equipment
10. Lighting Apparatus (not including the lighting apparatus with the voltage lower than 36V)
11. Motor Vehicles and Safety Accessories
12. Tire Products
13. Safety Glasses
14. Agricultural Machinery
15. Rubber Condoms
16. Telecommunication Terminal Equipment

[5] Catalogue of Compulsory Products Certification, http://www.cnca.gov.cn/cnca/rdht/qzxcprz/rzml/images/20080701/4755.htm

17. Medical Devices
18. Fire Fighting Equipment
19. Safety Protection Products
20. Wireless LAN Products
21. Home Decor and Remodeling Products
22. Toy Products

The Benefits of a CCC Mark

The benefits for a company that has undergone CCC certification for its products could be briefly summarized as follows:

- ✓ The company can freely harness the huge potential of the Chinese market and its products imported through Chinese customs without problems.
- ✓ After a successful first-time certification, new products can be certified quickly and inexpensively, saving the company money and time when submitting applications for additional products.
- ✓ More efficient quotes for CCC certified products, as a certified company has exact knowledge of costs and can quickly calculate offers accordingly.

Fines and Punishments

Through actions such as the unauthorized supply, sale, import or marking of products listed in the CCC Catalogue before they are certified, an offense is committed. Offenders will be charged with penalty fees ranging between 50,000 - 200,000 Chinese Yuan. Additionally, revenue generated through the sale of uncertified goods may be confiscated and nationalized.

4. When Certification is Unnecessary

General Problems with Customs

It frequently happens that the Chinese customs authorities demand a CCC certificate for a product that actually does not require one. This usually occurs when the customs code (HS code) or the name of the product is indicates that CCC is required.

In these cases, an exemption does not need to be applied for as the product does not need a CCC mark (and subsequently cannot get one, as certification is only possible for products that are listed in the Catalogue). In such cases, the importer should provide detailed product information to the local China Entry-Exit Inspection and Quarantine Bureau to receive a CIQ declaration proving that the product does not need CCC. When this document is provided to the customs officer in charge, customs clearance can proceed.

CCC Exemptions (CNCA Announcement No. 3, 2005)[6]:

In 2005, CNCA published Announcement No. 3 to regulate the import of products that are subject to CCC without being certified.

This exemption must be applied by the importer and, in general, can only be used for small quantities. The application must be submitted for every shipment and product tests can be requested.

---

[6] http://www.cnca.gov.cn/cnca/rdht/qzxcprz/myblrzzm/4734.shtml

There are several cases where an application for exemption can be submitted, most notably: products that are intended for research or testing in China, parts that are used to build a production line, spare parts that are required by the end-user, products used for fairs or trade shows, temporary import, and parts that are built into a whole product/machine that is intended for eventual export. In each case, the importer has to provide detailed documents explaining the reasons for the exemption and detailed product information as well as test reports.

Applying for an exemption is especially useful in the case of a small quantity of industrial spare parts that require a CCC mark and will be used for maintenance or for building a production line in China.

Small Volume Imports (CNCA Announcement No. 38, 2008)[7]:
Generally, products that require a CCC mark can be imported into China without a certificate through the small volume regulation as described in CNCA Announcement No. 38 from 2008. The importer must apply for a small volume import and customs can demand to perform tests according to Chinese GB standards for each shipment. The conditions and regulations vary depending on the product and port of import. Usually, using the small volume regulation for import only makes sense for valuable goods such as luxury cars, as the procedure takes time and testing can generate significant costs.

---

[7] http://www.cnca.gov.cn/cnca/zwxx/ggxx/66801.shtml

5. Relevant Authorities and Institutions

Functions of Various Authorities and Institutions involved in the Certification Process:[8]

(1) AQSIQ

The General Administration of Quality Supervision, Inspection and Quarantine of the People's Republic of China (AQSIQ, http://www.aqsiq.gov.cn/), is a ministerial administrative body directly subordinate to the State Council of the People's Republic of China, and is in charge of the nation's standardization, metrology, quality supervision, inspection and quarantine. Its functions in the China Compulsory Certification process are:

- To approve and release the Catalogue together with the Certification and Accreditation Administration of the People's Republic of China;
- To organize the rules and regulations for the China Compulsory Certification.

(2) CNCA

The Certification and Accreditation Administration of the People's Republic of China (CNCA, http://www.cnca.gov.cn), is the administrative body in charge of nationwide product certification authorized by the State Council of the People's Republic of China. Its functions in the China Compulsory Certification process are:

---

[8] http://www.cnca.gov.cn/cnca/rdht/qzxcprz/qzxcpzdjs/4713.shtml

- To draft and adjust the Catalogue, and to release the Catalogue together with the General Administration of Quality Supervision, Inspection and Quarantine of the People's Republic of China;
- To specify and release the implementation rules for the certification of products included in the Catalogue;
- To draft and issue the certification mark, and to define the requirements for CCC certificates;
- To designate the institutions responsible for certification tasks, the inspection authorities responsible for product testing and factory inspection, as well as the institutions responsible for issuing the CCC mark ;
- To publish certified products and manufacturers;
- To guide the investigation and treatment of illegal certification activities;
- To accept and hear complaints and accusations as well as organize the consequent investigation and punishment for serious illegal certification activities;
- And to approve the exemption of certification for special products.

(3) CQC

The China Quality Certification Centre (CQC, http://www.cqc.com.cn) is a professional certification body approved by AQSIQ and CNCA, and is designated by CNCA to process CCC certifications.

(4) CCAP

The China Certification Centre for Automotive Products (CCAP, http://www.cccap.org.cn/) is a certification body designated by CNCA that specializes in the certification of automotive products.

A Brief Introduction to the Authorities and Institutions involved in the CCC Process

General Administration of Quality Supervision, Inspection and Quarantine of the People's Republic of China (AQSIQ)

The General Administration of Quality Supervision, Inspection and Quarantine of the People's Republic of China (AQSIQ, http://www.aqsiq.gov.cn/) is in charge of national quality, metrology, entry-exit commodity inspection, entry-exit health quarantine, entry-exit animal and plant quarantine, import-export food safety, certification and accreditation, standardization, as well as administrative law-enforcement.

AQSIQ has 19 in-house departments including the General Office, Department of Legislation, Department of Quality Management, Department of Metrology, Department of Inspection and Quarantine Clearance, Department of Supervision on Health Quarantine, Department of Supervision on Animal and Plant Quarantine, Department of Supervision on Inspection, Bureau of Import and Export Food Safety, Bureau of Special Equipment Safety

Supervision, Department of Supervision on Product Quality, Department of Supervision on Food Production, Department of Law Enforcement and Supervision (AQSIQ Office of Fight against Counterfeits), Department of International Cooperation (WTO Affairs Office), Department of Science and Technology, Department of Personnel, Department of Planning and Finance, Party Committee (PC) Office and Bureau of Retiree Cadres, In addition, assigned by the Central Commission for Discipline Inspection of CPC and the Ministry of Supervision, a Discipline Inspection Team and the Inspection Bureau resident permanently in AQSIQ.

AQSIQ administers both the Certification and Accreditation Administration of the People's Republic of China (CNCA) and the Standardization Administration of the People's Republic of China (SAC). CNCA is a vice-ministerial-level department, exercising the administrative responsibilities of unified management as well as the supervision and coordination of certification and accreditation activities across the country. SAC, which is also a vice-ministerial-level department, performs nationwide administrative responsibilities and carries out unified management for standardization across the country.[9]

[9] http://english.aqsiq.gov.cn/AboutAQSIQ/Mission/

Certification and Accreditation Administration of the People's Republic of China (CNCA)

The Certification and Accreditation Administration of the People's Republic of China (CNCA, http://www.cnca.gov.cn), was founded to centrally administrate, supervise and comprehensively coordinate certification and accreditation nationwide, following a decision of the State Council of the People's Republic of China (http://www.gov.cn/) in April 2001. The decision was made to establish the General Administration of Quality Supervision, Inspection and Quarantine of the People's Republic of China (AQSIQ, http://www.aqsiq.gov.cn/) and to merge the Entry-Exit Inspection and Quarantine Bureau and the former Administration of Quality and Technology Supervision, and to establish the CNCA in the meantime. [10]

The China Quality Certification Centre (CQC)

The China Quality Certification Centre (CQC, http://www.cqc.com.cn) is a professional certification body under the China Certification & Inspection Group (CCIC, http://www.ccic.com) approved by the State General Administration for Quality Supervision and Inspection and Quarantine and the Certification and Accreditation Administration of the People's Republic of China. As the largest professional certification body in

[10] pp19, Encyclopedia of China Compulsory Certification,CNCA, CHINA METROLOGY PUBLISHING HOUSE

China, CQC evolved from the former China Commission for Conformity Certification of Electrical Equipment, which was established in 1985. In April 2002, CQC was created through the merger of six institutions under five different ministries (including the former China National Import & Export Commodities Inspection Corporation Quality Certification Centre; the Secretariat, the Electrical Equipment Subcommittee, the Home Appliance Subcommittee, and the Electronics Subcommittee of the former China Commission for Conformity Certification of Electrical Equipment; as well as the CCIB Beijing Review Office). In September 2007, a restructuring program was launched and CQC therefore became the dedicated certification platform of the newly incorporated CCIC.

The core business of CQC involves product certification—including China Compulsory Certification (CCC)—voluntary certification, management system certification and certification training services. Additionally, CQC is also a State authorized third-party certification body for the certification of energy-saving, water-saving and environmental-friendly products.

CQC is a national certification body (NCB) for the IEC System for Conformity Testing and Certification of Electrotechnical Equipment and Components and an official member of China in IQNet, IFOAM, ANF and CITA. CQC has been officially authorized by the Japanese government to offer compulsory product certification (PSE) service for Japanese products. It has also been authorized by

Germany-based KBA to provide Chinese applicants for E1/e1-Mark certification with factory inspection service. Additionally, the Chinese Ministry of Commerce has authorized the CQC to serve as the Export Commodities Technical Service Center. It has furthermore has established cooperative relations with 27 certification bodies from 19 countries and regions.

Dedicated to promote the economic and social development of China, CQC is now heading towards the goal of becoming an international certification body with high social credibility, strong innovation capability, market competitiveness, and sustainable development capability.[11]

China Certification Centre for Automotive Products (CCAP)

China Certification Centre for Automotive Products (CCAP, http://www.cccap.org.cn/) was established in August 1998. As the Automotive Product Certification and Management System Certification Body, it carries an impartial third-party position.

On Oct 16, 2002, CCAP registered as China Certification Centre for Automotive Products and obtained its business license from the State General Administration for Industry and Commerce to offer certification. In Dec 2003, CNCA issued the Approved Certificate for Certification Body to CCAP for quality management system certification and products certification.

---

11 http://www.cqc.com.cn/english/aboutcqc/CQCIntroduction/A022901index_1.htm

The Certificate of Accreditation of Certification Body for Products and Certificate of Accreditation of Certification Body for QMS have been issued to CCAP by CNAS. CCAP carries out the current international certification system and strictly abides by the regulations of state rules and laws related to the certification of products and quality management systems.

The aim of CCAP is to promote high quality automotive products as well as to serve society by ensuring safety, environmental protection, energy efficiency and quality through objective, fair and legal certification and scientific management.

To carry out its mandate, CCAP commands a team of professional auditors who were trained in state-designated institutions and passed qualifying examinations. CCAP also owns a couple of CNAS-accredited test labs that are capable of testing automotive products for certification purposes.[12]

[12] http://www.cccap.org.cn/eccap/AboutUs/Introduction.jsp

Chapter Review

The CCC certification system came into effect in 2002.

If a product is listed in the Catalogue, it requires CCC certification—whether imported or produced domestically.

The two regulating bodies for CCC certification are AQSIQ and CNCA.

Upon successful certification, annual follow-up certifications are required.

For the most accurate list of products requiring a CCC mark, both the Catalogue and announcements published by CNCA should be referenced.

Exemptions for importing products subject to CCC certification exist, such as for small volume imports, the relevant regulations for which can be found in CNCA announcements.

Chapter Two

Components of CCC

1. Application Procedure

First, it is important to check the product catalogue and announcements made by CNCA and AQSIQ to see if certification is compulsory for a specific product. If the product is listed in the Catalogue, the applicant will need to identify the appropriate certification body, by locating the list of approved DCBs published by CNCA[13]. Typically, the certification authority will be CCAP or CQC. Once the certification body is identified, the application may begin.

Application Requirements[14]

Domestic or foreign enterprises can apply for compulsory products certification. To do so, the following requirements should be satisfied:

(1) If the applicant is a domestic company, they must hold a Business License of Enterprise Corporation issued by the Administration for Industry and Commerce Organization; foreign enterprises must produce a Registration License issued by an equivalent entity from their country.

(2) The products requiring certification should conform to Chinese State Standards, Industrial Standards or other Additional Technical Requirements.

---

13 http://www.cnca.gov.cn/cnca/rdht/qzxcprz/jcjggljg/4731.html

14 http://www.cccap.org.cn/eccap/ApplicationGuide/ApplicationGuide.jsp

(3) Proof must be shown that the products requiring certification can be produced in batches with stable quality.

(4) The quality management system of the applying enterprise should conform to the Standard of series 2000 Edition ISO-9000 or other Quality System Standards (such as QS9000, GB16949 or VDA 6.1).

How to Apply[15]

Nowadays, most applications can be filed and accepted online but the application procedure itself might vary, depending on the designated certification body (DCB).

In most cases,

(1) Applicants who satisfy the basic requirements may submit the application to the DCB.

(2) The DCB will decide whether or not to accept the application within a couple of days after receiving the completed materials. If the application conforms to the requirements, a unique application number will be assigned and an Acceptance Notice will be sent by the DCB; otherwise, a notice informing the applicant of the reasons for the rejection will be sent.

[15] http://www.cccap.org.cn/eccap/ApplicationGuide/ApplicationGuide.jsp

(3) Upon approval, the applicant will the need to pay all certification-related fees.

(4) The DCB will work out the certification plan and start the formal certification process.

During the certification process, the applicant can check the progress through an online certification administration system for updates on the acceptance notice, sample delivery notice, factory inspection notice, charge notice and certificate issuance notice.

2. Product Tests

After an application has been accepted, the applicant should prepare the necessary documents as instructed in the notice issued by the DCB. These usually include a certification application form, the registration licenses of the applicant and manufacturer, a product description, a factory inspection questionnaire, a declaration of consistency, related qualification certifications of the sample products to be tested, a trademark registration if applicable, the assignment of quality managers as well as any related assessment certificates.
The testing laboratory in China will be assigned to the applicant by the certification authority, but it is also possible to request a specific lab, if it is a CNCA approved facility in China.

When informed to send samples for testing, the applicant should send the products, in the amount requested, along with the prepared documents as noted above to the designated testing institution in China. Problems can occur during the import procedure, as these are products that required a CCC certificate for import. Therefore, it is advised to use experienced logistic companies and import agents.

The testing typically takes 2-8 weeks, depending on the product. During certification, one product test for each application must be performed. The tests are performed according to Chinese GB (Guobiao, or, in English, "National Standard") standards.

While the Chinese standards are tantamount to European ECE and other international standards, Chinese authorities do not recognize any tests except those conducted in Chinese test labs for the purpose of certification.

All tests reports are issued in Chinese only.

3. The Factory Audit

The purpose of the audit is to ensure that Chinese standards for quality management are met and all relevant CCC regulations are known and fulfilled. The inspection is conducted with the aid of checklists and, in many aspects, is similar to other international quality management audits.

The CCC audits are to be taken seriously. Since the establishment of CCC in 2002, the auditors have gained considerable experience and the quality of the certification can compare with other international quality certifications. Thus, quality control will be scrutinized diligently for weaknesses. For specific divisions of a company that have received ISO-certification for their quality management system, the factory inspection will be significantly easier to pass.

The factory audit typically causes concerns for new applicants. However, the Chinese authorities are only checking the following four conditions:

- Does the quality system of the manufacturer follow Chinese requirements?
- Do the products presented for testing resemble the real products?
- How does the manufacturer ensure CoP (Conformity of Production) for its products?

- Are test results and quality checks properly documented?

Preparation for the audit should be carried out according to "Implementation Rules for Compulsory Certification". Updating any necessary documentation is also strongly recommended. In particular, Chinese auditors expect that CCC standards are explicitly recognized in the internal quality manuals of the plant. Additionally, descriptions of major manufacturing and inspection processes as well as existing approvals and official certifications should be compiled beforehand.

Before an audit takes place, all applicants should ensure the following:

- The quality control management fulfills the requirements of the Chinese certification authority (CNCA) and is updated regularly;
- The products produced during the audit are the same as the products sent to the test lab;
- The CCC marking process is recorded for any products that have already been granted certification and a documented procedure exists to ensure that only products that have received CCC certification are marked as such;
- The factory has test devices for the product or can ensure that regular external testing is done;
- All employees working in the area of quality management have relevant qualifications;

- And that the requirements of the corresponding GB standards are fulfilled.

During the course of an inspection, the manufacturer is obliged to grant the auditors access to all areas of production and quality management that are related to the parts being considered for certification. Access to other areas can be denied in a friendly manner.

The duration of the initial factory inspection can last anywhere between two and five days. For example, in the case of motor vehicle products, usually a team of two inspectors can conduct their audit within two days. In this case, one auditor may handle documentation while the other focuses on quality management or production. In most cases, the auditors will come from two different departments to ensure a more diligent audit.

The audit is the most costly part of the whole certification process, as highly qualified auditors must be paid and their travel and accommodation expenses covered for the duration of their trip. Additionally, the services of an interpreter may also be required, as auditors often do not speak sufficient English to conduct a smooth audit.

Since inspections are required for all plants where final production takes place, the inspection of several plants could be combined in an "inspection tour". This reduces the travel expenses and shortens the process.

One final important point to note in preparation: the inspection tour itself should be carefully organized to harmonize with the planning of the certification authority, taking into consideration all travel arrangements and corresponding visas. Organizing visas for Chinese citizens can be a lengthy process and it is important to plan accordingly; therefore, the dates for inspections should be agreed as early as possible.

4. The CCC Certificate

After evaluating the results of the product testing and factory inspections, the CCC certificate will then be issued or denied (with an explanation as to why). The certificate will contain the name of the applicant, manufacturer and plant, as well as trademarks, part numbers and product name, relevant Chinese norms and standards and the date of issue.

The CCC certificate will be issued in both English and Chinese and sent to the manufacturer. A corresponding entry is made in CNCA's database allowing customs officials to verify whether or not the imported goods are certified. Nonetheless, upon certification, the CCC mark must be affixed to every product before entering China.

5. The CCC Mark

After CNCA issues the "Permission of Printing", the CCC mark and factory code of the plant must be attached to the certified parts. For this, alternative methods are possible, but it is important to note that not all methods are approved for all components.

Due to differences in testing criteria, there are four distinct types of CCC marks. The following marks are for safety (S), electromagnetic compatibility (EMC), electromagnetic compatibility and safety (S&E) and fire protection (F).

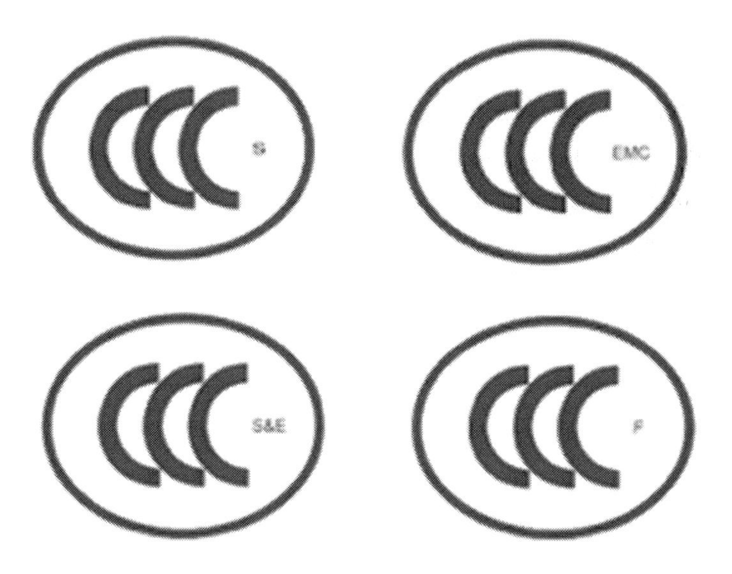

The Four CCC Marks

The most popular methods for marking with the CCC logo and factory code are:

• Pressure stamping, molding or laser-etching;

• Printing;

• Applying original CCC stickers (purchased from a CNCA-designated agent) or gluing a manufacturer-printed label to the component.

Products may only be labeled with the CCC mark and factory code after the permission of printing is given for that specific marking solution. Marking a product prior to permission involves high risks that may result in termination of the CCC certificate and additional fines.

The manufacturer can choose the size of the CCC mark, as long as it is within a designated range. However, the CCC mark and the factory code must always remain legible. Additionally, the proportions of the CCC logo may not be changed. On an original CCC sticker, the factory code is printed on the back of each sticker but for self-printed labels or other marking solutions, the factory code can be placed anywhere in the immediate vicinity of the logo.

Generally, the CCC mark for installation parts (in vehicles, machines etc.) does not have to be visible when the part is fitted in. Therefore, the CCC mark on design-related components will usually be hidden on the back or underside. Exceptions include specific parts such as headlights, reflectors and other selected products. For these parts, the CCC mark must be clearly visible when installed. For all other products, such as IT products, the CCC mark and factory code are usually integrated on the nameplate.

The Permission of Printing is only valid for one year and must be extended by the manufacturer. This can usually only be done if the factory participates in the yearly follow-up audits.

6. Annual Follow-Up Certification

Follow-up inspections must be initiated by the manufacturer. Thus, manufacturers should be conscious of time, contacting the Chinese authorities to conduct a follow-up audit before the CCC certificate becomes invalid.

Significant differences in the complexity and extent of the fields to be examined exist between the initial inspection and follow-up inspections, In most cases, the initial inspection is carried out in about two days by two inspectors, but it usually only takes one day in a follow-up audit. The follow-up inspections will take place approximately every 12 months after the initial audit and must be performed every year.

If too many flaws are found during any of the annual follow-up audits, the auditors could require a full inspection.

Chapter Review

Once you are sure that your product needs CCC, locate a CNCA approved Chinese authority to file an application.

Products for sample testing should be sent to the designated testing institution in China along with any necessary documents, as instructed.

Factory inspection is required for all plants where final production takes place.

The factory audit (both initial and follow-up) should be taken seriously and prepared for thoroughly.

Upon certification, choose the CCC marking most suitable for your product and which conforms to regulations.

Chapter Three

Additional Considerations

1. Fees and Cost Control

CCC certification is not expensive compared to other international product certifications.

The fees charged by CNCA usually include[16]:

(1) An application fee. The application fee will be charged upon the submission of the certification application. The application fee, as of 2013, is 500 Yuan.

(2) An application translation fee. If the applicant's application documents are not in Chinese, a translation fee will be charged. The exact amount of translation fee may vary, but will not exceed 1,000 Yuan.

(3) A testing fee. The testing fee is charged for product testing and the issuance of test reports, based on relevant product certification standards and technique requirements.

(4) A factory inspection fee. The certification authority sends out staff to execute document checking and factory auditing according to relevant factory inspection requirements.

The factory inspection fee varies with the number of auditing staff and days required for the audit. Typically, the price for an auditor is 2500 Yuan per day, but this number may be higher depending on a number of factors and significant price increases are expected for the coming years. The

[16] C.f. CNCA Announcement: No. 20, 2009

certification body determines the number of staff and days necessary for the audit, but typically, two staff will be sent for a factory audit of an automotive supplier, which will last two days.

Travelling costs of the audit staff and accompanying translator are also to be covered by the applicant.

(5) Approval and registration fee. Once a product has been approved, a CCC certificate will be granted and an approval and registration fee of 800 Yuan will be charged. For each approved application, a separate approval and registration fee will be levied.

(6) Supervision and re-certification fee. For any new application the initial application fees will be charged except the audit fees.

(7) Annual fees. Annual fees are charged at a rate of 100 Yuan for each CCC certificate

(8) Marking fee. The fee charged for one permission of printing is 450 Yuan.

Regulations regarding pricing standards:

- CNCA Announcement: No. 20, 2009

- NDRC Price [2006] No. 179

For illustrative purposes, the following is an informal example of the certification costs for two vehicle headlights with three different functions each (two applications). Please note: this example does not include consulting fees or the cost of a translator.

EUR 1 = CNY	8,0
USD 1 = CNY	6,3
EUR 1 = USD	1,3
Amount of Certification Units	2
Test fees for one Certification Unit in CNY	15000

Description	Price	Currency	Factor	Total	Total in EUR	Total in USD
Test Fees	15000	CNY	2	30000	3750	4800
Application Fee	500	CNY	2	1000	125	160
Application Translation Fee	1000	CNY	2	2000	250	320
Approval and Registration Fee	800	CNY	2	1600	200	256
Other charges by regulation	220	CNY	2	440	55	70
Marking approval Fee of CNCA (Printing Permission)	450	CNY	1	450	56	72
Factory Inspection Fee	18000	CNY	1	18000	2250	2880
Travel fees of auditors international	19000	CNY	1	19000	2375	3040
Accomodation expenses of auditors	800	EUR	1	800	800	1040
Travel fees and expenses of auditors domestically	780	EUR	1	780	780	1014
Total					**10641**	**13652**

Besides the fees charged by the certification body, an applicant should also take into account other costs that may be incurred during the certification application process, such as:

(1) Costs for interpreter/translator(s) that assist in communicating with the Chinese auditors and plant staff during a factory inspection;

(2) The costs of products used for testing as well as their shipment to test laboratories in China;

(3) Work hours of plant staff;

(4) And any consulting fees for a professional CCC service agent.

2. Managing Changes

The CCC certificate contains a reference to the underlying test specification (GB standard and implementation rule) and information about the manufacturer and certified products. The importance of this information is often not sufficiently acknowledged as the certificate becomes invalid if any of the following details change:

- The name of the manufacturer,
- The address of the manufacturer,
- Relocation or expansion of production into a new manufacturing plant,
- New model name is given to the product.

[Please note: if a part number / item number was specified as the model name in the application, then even product changes that are not actually relevant for certification, such as color changes, can make a CCC recertification suddenly necessary, just because the part number has changed.]

Recertification is also required if any of the other information on the CCC-certificate changes, such as:

- A change is made to the product itself,
- A change of supplier,

- CNCA has changed the implementation rules or standards (for example, they issue a newer version of a GB Standard).

Also, any change in the form and style of CCC marking requires a new printing permission from CNCA.

3. Frequently Asked Questions about CCC

How do I find out whether my product requires CCC?

A list of products as well as regular announcements about new products that require certification are published by AQSIQ and CNCA.

Do I need to travel to China during the CCC process?

No, it is not necessary to travel to China. Sometimes, however, it can be useful to send someone to coordinate product testing with the test labs, which can simplify the procedure, lead to fewer tests and thereby cut costs. Usually, it is advisable to hire a professional CCC agent for this.

Is it necessary to take care of cultural differences when dealing with the auditors?

Yes. Regular intercultural tips for dealing with Chinese people, especially government officials should always be followed.

No matter what happens during the audit, it is always wise to stay courteous and calm. It is advisable to deal professionally with all reasonable requests from the auditors. Say "No" only in extreme cases and work with the interpreter to choose the right words to deal with difficult matters.

Does the quality manager or anyone else need training?

Usually no special training is needed if the factory has ISO or any other quality management certificate. If this is not the case, then it may be necessary for the quality manager and possibly his/her team to receive some training to fulfill the certification requirements.

How can I prepare myself best for certification?

The tests necessary for CCC certification follow a certain GB Standard. Thus, it is useful to check the product standards, understand the Chinese regulations and read the corresponding checklists.

Will my certificate lose viability?

Your certificate will not lose viability if you do the yearly follow-up audits. There is an expiration date on most CCC certificates, which is usually four to five years after the date it was issued; thereafter, you will receive a new version upon re-inspection.

What do I do once I get CCC, can I just export the marked parts to China and expect hassle-free cross-border shipment?

Yes, with your CCC certificate, the Chinese markets are 100% open for you to enter and flourish.

Management of changes: What are the procedures? What should I do to optimize beforehand?

Every change of CCC certificates involves costs. If no tests are needed then only application fees need to be covered.

Manufacturers always need to seek a good management of changes. Either they study the rules themselves and apply them according to their industry or they take advantage of an experienced service provider to advise them accordingly. The automotive industry is a good example of an industry where good change management is important.

Where do I best obtain stickers? What do they cost?

You can buy original CCC stickers from a CNCA-designated agent in China. The costs for stickers are generally very low, between 0.01 – 0.08 USD.

Many people report CCC problems suddenly arising after years of hassle-free cross-border shipment. How is this possible?

There is no protection against sudden changes. Rules can change--sometimes with only three months notice (although 12 months is the typical announcement time). And as announcements are not made in English, this can complicate matters further.

Additionally, customs officers can make mistakes and assume a product needs certification when it does not. This also causes interruptions.

I heard about a case where a quality manager was requested to travel to China to correct the position of a CCC-sticker. Is this just a rumor or are the Chinese customs really so prone to cause problems?

This story is—as far as we know—only a rumor, though rumors sometimes contain truths. Possibly, there was a condition that fostered this outcome. Things like this can happen for car manufacturers during harbor check, but usually customs officials do not deliberately scour for flaws if the documentation is correct.

Is it a good idea to hurry Chinese authorities?

As time often plays a crucial role, one is tempted to put pressure on the Chinese authorities to quicken the process. This should only be attempted if the applicant has personal, long-standing contacts. Without strong contacts, trying to hurry the Chinese authorities will only have a negative effect.

Is there an authority I can escalate problems to if I feel I am being mistreated by Chinese authorities?

Complaints can be made to CNCA, though this is only useful in very particular cases.

In automotive certification, when is an "all parts certification" not the best option?

The "all parts certification" might not be the best option when small volume series, such as sport cars, are to be exported to China. In this case, it is better to certify the whole car first and then decide later on this issue of service parts.

Will CCC still exist 10 years from now? Will it evolve?

There is no sign of CCC decreasing in scope in the near future; instead, it is being expanded continuously. We expect CCC to evolve with the passing of time.

What is the impact of economies of scale in the CCC-Certification?

The economies of scale are very significant. A well-planned certification for different plants owned by one company can enormously cut costs compared to random certification of single parts, plants or small series.

Chapter Review

CCC certification is not expensive, but all possible expenses should be taken into account when budgeting.

Before any alteration is made to a product, check if the CCC certificate will still be valid after the change.

Observe if the GB standard or the implementation rules have been updated to keep the CCC certificate valid.

Chapter Four

Automotive CCC

1. Special Features of the Automotive CCC

In the area of CCC certification for auto parts and vehicles, the number of Chinese regulations and requirements is particularly high. Deadlines must be strictly observed, product tests and audits are very thorough and the change management ought to be strategically prepared for, given the dynamic nature of the automotive sector.

In addition to clarifying if certification is mandatory for specific products and summarizing the products in the certification units, it is important to identify which factory will be audited. Typically, the factory where either the total product is made or the last quality assurance steps are taken is the one to be audited.

The car itself, as well as several individual parts, is subject to CCC. Especially when it comes to service parts, a wise choice regarding parts certification needs to be made, where costs and potential customs problems should be taken into account.

2. Practical Instructions for Certification

Test labs

The test laboratory for product tests in China will be assigned to the applicant by the certification authority, but it is also possible to request a favored laboratory in China.

Whole Car Certification v. Parts Certification

The car itself is subject to certification, and numerous individual components within are subject to CCC.

Choosing to certify individual parts is the preferred solution, otherwise service parts imported to China may be detained by Chinese customs and exemptions will need to be applied for. For the homologation of the whole car, it is not required that every component has an individual CCC certificate from the supplier, but homologating the car can also mean a lot of effort when changes to the components or suppliers are made. Additionally, as individual components cannot be marked with the CCC mark shipment of service parts to China will be a problem if the supplier does not have an own CCC certificate.

However, the "all parts certification" might not be the best option when it comes to small volume series export, for example, sports cars. In such cases, it might be better to certify the whole vehicle first and the shipping of service parts later on or certify only the most important service parts.

Part Numbers and CCC

The specification of the vehicle manufacturer's part number on the CCC should be avoided in most cases, as these part numbers could change due to simple modifications that are not relevant to the certification (e.g. change in drawings). In such cases, the CCC certificate would have to be changed, which causes unnecessary effort and expenses.

It is better to use a supplier's part number or unique part name. Of course, this number/ name must also be implemented on the product.

Marking

The manufacturer of automotive parts can choose a type of CCC marking from the following options:

- Applying original CCC stickers purchased from an official agent,
- Gluing a self-printed label on the component,
- Pressure stamping, molding or laser-etching,
- Printing directly on the component.

The original stickers are inexpensive and the certification authority keeps a clear record of the amount of stickers purchased. The use and management of stickers should be strictly documented to guarantee that no such stickers are applied to products that are not certified. As this requires a great deal of effort both to attach the stickers as well as documenting them, it is not the recommended way of marking for any production line other than small volume output products such as parts for super luxury cars.

Also, it is worth repeating again that only after the CCC certificate and the printing permission are granted can the products be CCC marked.

Additionally, to avoid problems that occur when the names/part numbers on the CCC are not an exact match to the names/part numbers on the product, it is better to integrate names on the parts or self-printed labels.

Documentation

An update of documentation is part of the preparation for a factory audit.

Ideally, the CCC standards should be explicitly recognizable in any internal quality manuals; this can be done, for example, by mentioning the Chinese testing rules in the internal processes.

Additionally, the CCC marking process needs to be documented to ensure that no products are marked without accompanying certification.

Good Communication with Auditors

Especially in the field of automotive CCC, know-how plays an important role. The translator should be informed to alert the auditors whenever they perform activities that are beyond their authorization (such as taking photos, requesting confidential documents or downloading items from computers). Since there is growing sensitivity to the problem—also on the Chinese side—usually no offense is taken.

3. Interior Trim Parts

It is easy to check if CCC is required for products such as horns, outdoor lighting, and fuel tanks, but can be more complicated for interior parts. Certification is required for parts such as door panels, pillar trim and inner linings of a coupe. But, as of the beginning of 2013, products such as removable floor mats, parts of the instrument panel and parts in a closed trunk are not required to be certified.

For interior trim parts, the relevant standard is GB 8410 (from 2006) and the relevant implementation rule is CNCA-02C-060: 2005.

All seat trim covers and interior trimming parts that have a size smaller than 356mm x 100mm are no longer subject to CCC certification according to Announcement No. 117 [2012] from CNCA[17].

For the following components, CCC requirements should be cleared with the certification authority (particularly if a service part delivery is planned): parts of the center console, parts in the footwall area, decorative trim parts and parts of a door panel.

When it comes to product testing for interior parts, only the flammability of products will be tested.

Parts of different sizes can be combined in one CCC certification unit, if the following conditions are met:

- There is only supplier for each material and only one production site;
- Parts contain the exact same materials;
- Parts have the same thickness for each layer.

Please note: colors do not matter if the materials and material types are all the same.

[17] Announcement No. 117 [2012] from CNCA, Source: http://www.cnca.gov.cn/cnca/zwxx/ggxx/648334.shtml

The position of the CCC mark should be carefully chosen along with the printing process by which it is applied. Ideally, the logo should appear in a place where it is invisible in the installed state of the component (in interior trim parts, the CCC logo does not have to be visible).

4. GB Standards

The GB Standard (GB stands for *Guobiao*, or "National Standard") is the basis of testing for products that require certification. Therefore, if there is no corresponding GB Standard for a product, CCC is not required.

GB standards are issued by the Standardization Administration of China (SAC). SAC is the institute that represents the Peoples' Republic of China in the ISO and IEC. As the name implies, GB standards are the national standards; "GB" is prefixed to mandatory national standards and "GB/T" is prefixed to recommended national standards, (the 'T' stands for Tuijian, Chinese for "Recommendation").

The Chinese standards correspond largely to the European ECE and other international standards. Nevertheless, for the purpose of certification, the Chinese authorities do not recognize any tests except those conducted in designated Chinese test labs.

For complete implementation of the certification of the entire vehicle as well as to ensure a smooth flow of supplies for service parts, it is necessary to certify products in the following product areas:

Product Name	CNCA-Rule	GB Standard	Corresponding International Standard
Seat belt	CNCA-02C-026	GB 14166	ECE R16
		GB 8410	FMVSS571.302
Tire	CNCA-02C-027	GB 9746 GB/T-2978	
Safety glass products	CNCA-02C-028	GB 9656	ECE R43
Headlamp	CNCA-02C-058	GB 4599 GB 4660 GB 11554 GB 5920 GB 15235	ECE R6 ECE R7 ECE R19 ECE R38
Front position lamp/ rear position lamp, parking lamp, outline marker lamp, braking lamp	CNCA-02C-058		

Turning-signal lamp	CNCA-02C-058	GB 17509 GB 18409 GB 18099 GB 18408 GB 25991	
Reversing lamp	CNCA-02C-058		
Front fog lamp, rear fog lamp	CNCA-02C-058		
Rear license-plate light	CNCA-02C-058		
Side-marker lamp	CNCA-02C-058		
Retro reflector	CNCA-02C-056	GB 11564	ECE R3
Seat and head restraints	CNCA-02C-063	GB 15083 GB 11550 GB 13057 GB 8410	ECE R17 EEC 78/932 ECE R80 FMVSS571.302

Brake hose	CNCA-02C-057	GB 16897	FMVSS 106
Rear-view mirror	CNCA-02C-059	GB 15084	ECE R46
Fuel tank	CNCA-02C-062	GB 18296	ECE R34
Horn	CNCA-02C-055	GB 15742	ECE R28
Interior trimming material: floor covering, seat shield, decorating scale boards (inside door shield/panel, front wall inner shield, side wall inner shield, rear wall inner shield, roof liner)	CNCA-02C-060 CNCA Announcement No. 117 [2012]	GB 8410	FMVSS571.302
Door lock and door hinge	CNCA-02C-061	GB 15086	ECE R11

Source: scope of certification
http://www.cqc.com.cn/chinese/cprz/CCCcprz/rzfw/webinfo/2009/12/1260325225327502.htm

5. Classifications

Implementation rules are the regulatory documents issued by CNCA regarding all products subject to CCC. These documents define, in detail, all requirements for certification, including application units, type tests, factory inspection, accreditation, post-certification supervision, application of the CCC marks as well as changes and extension of certification. The implementation rules are the foundational documents used for the certifying authority to implement certification, for the applicant to apply for certification as well as for the local law enforcing bodies to supervise specific products.

In total, there are 47 implementation rules issued by CNCA for products in different categories. Below is a list of all automobile products and their corresponding classification and implementation rules.

Category	Product Name	Scope	Implementation Rule
1101	Motor Vehicle Products	Motor vehicles of categories M, N and trailers of category O operating on the road in China. Chassis of category 3 (the chassis of trucks without cab and dash panel) and agriculture vehicles are not included	CNCA-02C-023
1104	Safety Belts	Safety belts	CNCA-02C-026
1106	Horn for Motor Vehicles	Horn for motor vehicles of M, N, L3, L4, L5 categories driven by direct current and compressed air	CNCA-02C-055
1107	Retro-reflector for Motor Vehicles	Retro-reflector used for motorcycles, mopeds, motor vehicles and trailers	CNCA-02C-056

Category	Product Name	Scope	Implementation Rule
1108	Brake Hose for Motor Vehicles	Hydraulic pressure, air pressure and vacuum brake hose assemblies used by vehicles, trailers, motorcycles and mopeds	CNCA-02C-057
1109	External Lighting and Light Signaling Devices for Vehicles	Various kinds of external lighting and light signaling devices used for vehicles of M, N and O categories, including headlamp, front fog lamp, rear fog lamp, position lamp, end-outline marker lamp, brake lamp, reversing lamp, direction-indicator, parking lamp, side-marker lamp and rear-registration plate illuminating device, but does not apply to retro-reflectors	CNCA-02C-058

Category	Product Name	Scope	Implementation Rule
1110	Rearview Mirror for Vehicles	Rearview mirrors for motor vehicles of categories M and N, and for other motor vehicles having less than four wheels with the cab partly or wholly closed by the body	CNCA-02C-059
1111	Interior Trimming Material	Interior trimming products made of single-mode or stratified composite organic materials for cabs and cabins, including: floor covering, seat shield and all the decorating scale boards (including inside door shield, front wall inner shield, side wall inner shield, rear wall inner shield, and roof liner), excluding seat trim covers and those parts that have a size smaller than the basic sample size (i.e. smaller than 356 mm x 100 mm)	CNCA-02C-060 CNCA Announcement No. 117 [2012]

Category	Product Name	Scope	Implementation Rule
1112	Door Lock and Door Retention Components	Door lock and door retention components on each door for passengers passing in and out for motor vehicles of M1 and N1 categories	CNCA-02C-061
1113	Fuel Tank for Motor Vehicles	The metal and plastic fuel tank products for motor vehicles of M and N categories fueled with gasoline and diesel oil	CNCA-02C-062
1114	Seat and Seat Headrest Products for Motor Vehicles	The seat products for motor vehicles of M and N categories and the outside headrest products for the front seats for motor vehicles of M1 category	CNCA-02C-063
1201	Tire Products	Radial and diagonal tires of passenger cars and trucks	CNCA-03C-027

Category	Product Name	Scope	Implementation Rule
1301	Safety Glass Products	All kinds of safety glass used in vehicles, including laminated glass A/B used as windscreens in vehicles, zone-toughened glass for windscreens, laminated glass A/B or toughened glass used for products other than vehicle windscreens	CNCA-04C-028

Chapter Review

In most cases, it is better to individually certify all parts of a car.

For the certificate, use a supplier's part number or unique part name that will not change.

Keep the interpreter informed during the audit.

Only the flammability of interior trim products will be tested.

Relevant GB standards are the basis for product testing.

Implementation rules define all requirements in details.

Chapter Five

Recommendations

1. Summary

All-Inclusive Certification

If you want to achieve your CCC approval fast and hassle-free, hiring an experienced CCC service firm can be invaluable.
Since 2005, China Certification has supported hundreds of manufacturers from a variety of industries with their CCC certification.
We understand that CCC can be a complex hurdle for manufacturers and therefore, we competently cater to their needs at every level.
We offer the following services:

1. Checking the necessity of certification for products;
2. Complete CCC application preparation (including assistance with form completion, communication with the certification authority, and handling the necessary payments);
3. Complete handling of the product testing (completion of the customs clearance, document exchange, handling of payments, communication with Chinese test labs and evaluation of test reports);
4. Complete audit preparation (application, coordination and communication with certification authority, document exchange, handling of payments, organization of inspection and travel program, as well as providing the required documents, information and checklists for the audit preparation);

5. Complete audit escort (travel escort for the auditors, professional interpreter (Chinese-English) for the audit and checking of the inspection report);
6. Complete preparation for obtaining the CCC printing permission (advice on marking best practices, application for printing permission and templates for product labeling with the CCC logo and factory code);
7. Complete organization of follow-up certification, including audit;
8. And prolongation of the permission of printing

If needed, we can also provide pre-audits to ensure a positive outcome of the CCC audit performed by Chinese auditors. We also assist our clients with revising their quality management documents as well as any other documented procedures related to the CCC certification.

China Certification maintains excellent long-standing relationships to key personnel of authorities and test labs in China. With subsidiaries in Beijing and Shanghai, we maintain a presence in China, which allows us to offer optimal service for all CCC applicants. We combine Western and Chinese strengths in our international management team, providing a multi-cultural understanding that shows in our work with clients, partners and certification authorities.

The headquarters of MPR GmbH – China Certification are based in

Frankfurt, Germany. With our Chicago-based subsidiary, China Certification Corporation, we also serve the needs of our clients in the USA, Canada and Mexico. Our subsidiaries in Shanghai and Beijing cover the Chinese market and support direct and efficient communication with the Chinese authorities and test labs.

2. Appendix

List of Abbreviations

ANF: The Asia Network Forum

APLAC: The Asia Pacific Laboratory Accreditation Cooperation

APQP: advanced production quality planning

AQSIQ: The Administration of Quality Supervision, Inspection and Quarantine

CAQP: The China Association for Quality Promotion

CCCAP: The China Certification Center for Automotive Products

CCC: The China Compulsory Certification

CCEE: The China Commission for Conformity Certification of Electrical Equipment

CCIB: The China Commodity Inspection Bureau

CCIC: CHINA CERTIFICATION & INSPECTION GROUP

CEMC: The China Certification Centre for Electromagnetic Compatibility

CIQ: The China Entry-Exit Inspection and Quarantine Bureau

CITA: The International Motor Vehicle Inspection Committee

CNAL: The China National Accreditation Board for Laboratories

CNAS: The China National Accreditation Service for Conformity Assessment

CNCA: The Certification and Accreditation Administration of the People's Republic of China

CoP: Conformity of Production

CPK: process capability index

CQC: The China Quality Certification Centre

CRBA: The China Registration Board for Auditors

EMC: electromagnetic compatibility

GB: Chinese for *Guobiao*, or "national standard"

HS: The Harmonized Commodity Description and Coding System

IEC: The International Electrotechnical Commission

IECEE: The IEC System for Conformity Testing and Certification of Electrotechnical Equipment and Components

IFOAM: The International Federation of Organic Agriculture Movements

ILAC: The International Laboratory Accreditation Cooperation

IQNet: The International Certification Network

ISO: The International Organization for Standardization

MOF: The Ministry of Finance of the People's Republic of China

MPS: The Ministry of Public Security of the People's Republic of China

NCB: national certification body

NDRC: The National Development and Reform Commission of the People's Republic of China

PPAP: production part approval process

QC: quality control

QMS: quality management system

SAC: The Standardization Administration of People's Republic of China

TMP: testing at manufacturers' premises

WMT: witnessed manufacturer's testing

Links

Certification and Accreditation Administration of the People's Republic of China http://www.cnca.gov.cn

China Quality Certification Centre http://www.cqc.com.cn

China Certification Center for Automotive Products
http://www.cccap.org.cn/

Standardization Administration of People's Republic of China
http://www.sac.gov.cn/

- **Sources / Resources**

Implementation rules issued by CNCA and other related regulations in English:
http://www.cnca.gov.cn/cnca/cncatest/20040420/column/227.htm

Online query platform for CCC verification:
http://www.cnca.gov.cn/cnca/cxzq/rzcx/114453.shtml

Contact

<table>
<tr><td>China Certification Corporation

Three First National Plaza
70 West Madison St., Suite 1400
Chicago
Illinois 60602-4270
USA</td><td>Tel: +1-773-6542673
Fax: +1-773-6542673

E-Mail:
usa@china-certification.com
Web:
www.china-certification.com/en</td></tr>
<tr><td>MPR GmbH

Kaiserstr. 65
D-60329 Frankfurt am Main
Germany</td><td>Tel: +49 69 271 37 69 13
Fax: +49 69 271 37 69 11

E-Mail:
info@china-certification.com
Web:
www.china-certification.com/en</td></tr>
<tr><td>Detang Germany Business Consulting Shanghai Co., Ltd.

1106A, China Fortune Tower
1568 Century Avenue, Pudong District
Shanghai 200122
P.R. China</td><td>Tel: +86 21 6163 2960-889
Fax: +86 21 6163 2969

E-Mail:
info@china-certification.com
Web:
www.china-certification.com/en</td></tr>
<tr><td>Beijing Detang Dingfeng International Investment Consulting Co., Ltd.

19th Floor, No. 6, Workers' Stadium Road N.
Chaoyang District
Beijing 100027
P.R. China</td><td>Tel.: +86-10-59752686
Fax: +86-10-59752755

E-Mail:
info@china-certification.com
Web:
www.china-certification.com/en</td></tr>
</table>

About the Author

Mr. Julian Busch

CEO of China Certification Corporation in Chicago and Managing director of MPR GmbH in Frankfurt am Main, Germany

Julian Busch is a CCC specialist and has assisted numerous manufacturers in Europe, the US and Mexico with their CCC certifications over the last eight years. Mr. Busch is co-founder of the Germany-based consulting firm MPR GmbH and US-based consulting company China Certification Corporation. Together with his professional team in Germany, the US and China, he provides full certification packages for major car manufacturers, automotive suppliers and industrial companies from market leaders to microbusinesses. During the course of his career, Mr. Busch has spent several years in China and speaks fluent Chinese.

Copyright Notice

All rights reserved by Julian Busch. No part of this publication may be reproduced in any form, by photocopying or by any electronic or mechanical means, including information storage or retrieval systems, without permission in writing from Julian Busch. The sharing and forwarding of this publication is permitted only if the document is transmitted in complete and unchanged form.

Disclaimer

This publication was created with care and diligence according to the current standards in place at the time of writing. However, Julian Busch make no representations or warranties with respect to the accuracy or completeness of the contents of this work. The advice and strategies contained herein may not be suitable for every situation.

Made in the USA
Las Vegas, NV
11 August 2021